Predicting Storms 101

Robert Ellis

Published in Australia in 2024 by Goldener-Parnell Publishing
Reprinted 2025
Reprinted 2026
Reprinted 2027

Email: rob@worldstormcentral.co
Website: http://www.worldstormcentral.co

ISBN 9780646704876 (paperback)

Disclaimer
The author has made every effort to ensure the accuracy of the information within this book was correct at the time of publication. The author does not assume and hereby disclaims any liability to any party for any loss, damage or disruption caused by errors or omissions, whether such errors or omissions result from accident, negligence, or any other cause.

Graphic design by Bronwyn Melville

This book is dedicated to the memory of my parents.

Contents

What is the best time to watch for storms?

The relative humidity for a normal thunderstorm before storm onset needs to be above 80%, except for a fire thunderstorm where it will plunge to 10% or less.

Moisture is the fuel of storms.

According to NASA "While over land, thunderstorms are most likely to occur at the warmest, most humid part of the day, which is usually the afternoon or evening. Over the ocean they are most likely to occur in the early hours of the morning before dawn."

Thunderstorm Cloud Sequence

Arrival of a sequence of different cloud types can indicate a weather change. Check your barograph for an accurate prediction of a storm. Always keep one eye on the barograph trace and the other eye on the sky. Cirrus clouds indicate the wind direction of the jetstream level winds, which are narrow bands of strong wind in the upper atmosphere.

The further things are away from you, the slower they appear to move. You would expect very distant cirrus clouds to appear almost stationary. So, if you see them moving, they must be going very fast and you can expect a strong change.

When you can see only cirrus clouds, expect fair weather for the next 12–24 hours.

1: Cirrus clouds at sea.
Credit: Lieutenant Elizabeth Crapo, NOAA Corps

When cirrostratus clouds immediately follow cirrus clouds, you can usually expect a storm or a snowstorm in 12–24 hours.

Cirrostratus Clouds

Reliable Storm Prediction Rules

You will learn here how to predict storms approaching your immediate location long before they are even visible to radar and satellite.
Storm prediction is epic when it is reliable.

Learning to predict storms is easy and fun. You will treasure having this skill for the rest of your life. You need to initially learn to predict and enjoy every storm. You need to first learn independently how to ***predict storms approaching your immediate location,*** including predicting Severe Thunderstorms.

The Lightning Threshold is exceeded and an ordinary thunderstorm onsets when the sea level pressure falls more than 3.0 mb over 3 hours (or less) to below 1009 mb.

This is called ***The Thunderstorm Rule*** that we will explain shortly.

If this storm threshold is passed without the storm onset, storm development will deepen until the eventual onset of a Severe Thunderstorm such as a Supercell Thunderstorm.

The inclusion of a pressure sensor on good smart phones was a ***breakthrough in predicting storms*** and a real game changer.

Apps such as *Marine Barograph* can now read accurate pressure values at your immediate location. Often forecasts from the various agencies are based pressure readings taken too far away from the storm's pressure centre and are consequently unable to reliably predict storms.

The simple rules given in this book when applied to the accurate real-time pressure values taken by your phone's pressure sensor at your immediate location are reliable. Anyone can predict storms using the pressure sensor and a barograph app on their smart phone.

The smart phone with a barometer sensor and a barograph is a most precious standalone device. It tells you of an approaching storm long before it is even visible to radar or satellite providing an ***early warning.*** It is not affected by poor reception or power outages. You can use it at sea and in remote areas where there is no radar coverage. It is more economical if it uses pressure sensors and does not require the internet. The rules given here are essential to interpret your observations.

See overleaf for info about getting started with your smart phone.

Get a Barograph and Get Started

Pressure is a key input to atmospheric computer models used to predict the weather. General weather is affected by many factors, including pressure, temperature, wind speed, and humidity.

Fortunately, a storm's maximum wind speed (intensity) depends entirely on its central pressure. That is why the barograph is such a powerful instrument for predicting storms.

You can know how storms are actually developing by watching the barograph trace on your smart phone.

Getting started:

1. Download MARINE BAROGRAPH app. (iPhone, iPad, iPod touch or Android). App can be used on land or at sea.

2. In the app, enter your Elevation (height above sea level) in metres or feet. Plenty of free apps give your elevation e.g. *Current Altitude.*

3. Go to 'Graph' in the menu at the bottom then tap on 3h which is just above the graph itself. You usually need the pressure change over 3 hours.

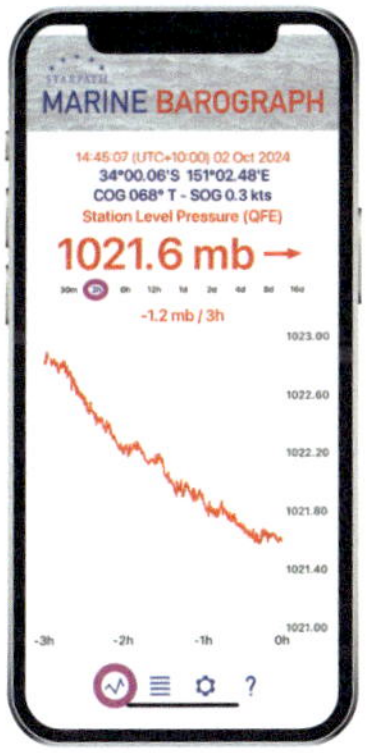

4. Open to page 3 of the book for ***The Thunderstorm Rule.*** Always keep your device still when monitoring the pressure. How much has the pressure reading shown on the barograph changed in the last hour? When the pressure starts falling at more than 1.0 mb / hour a storm may be approaching. Apply ***The Thunderstorm Rule*** to be certain.

5. Turn off Auto-Lock in your phone's Display Settings.

The Lightning Threshold is exceeded when the sea level ***pressure falls more than 3.0 mb over 3 hours (or less) to below 1009 mb (The Thunderstorm Rule).*** We can rely on this Rule because of its link to the Lightning Threshold and the laws of Physics on which we all depend.

The Wind Speed Threshold for the smallest storm is **25 km / hour**.
The smallest storm ever recorded, with a 3 mb fall in barometric pressure over 3 hours (1009 mb to 1006 mb), was recorded at Middlebury, Vermont, US on 7 June 2011. Del Genio *ibid* showed that the Lightning Threshold will be reached sooner and the updraft speed will increase by about 1 m/s if the atmospheric CO_2 content doubles.

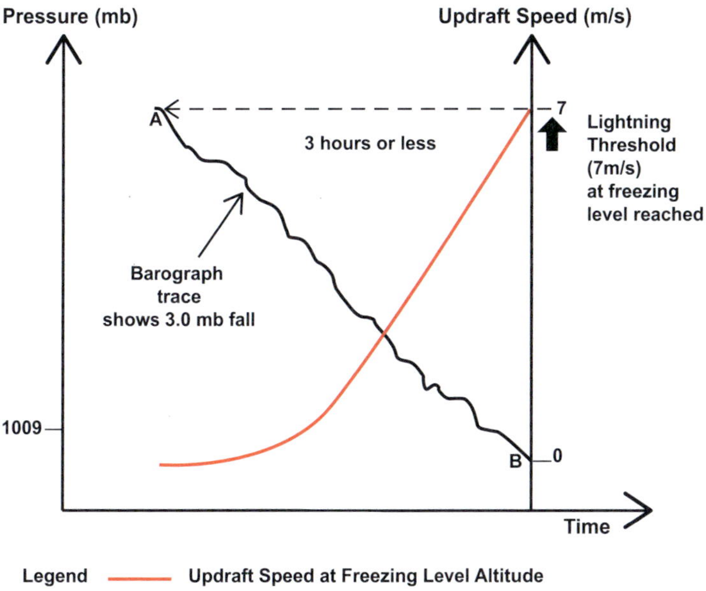

Figure 2. Lightning Threshold is reached when The Thunderstorm Rule is met

Thunderstorms can occur under certain conditions when the pressure is 1009 mb or above. This happens when localized areas of rising air reach the ***Lightning Threshold*** speed (7 m/s), near a Cold Front or a sea breeze. The lifting of the warm air by an advancing Cold Front causes intense and widespread thunderstorms. The storms occur in the localized areas of low pressure embedded in the Cold Front. This is a typical winter weather pattern.

When does a Storm Start?

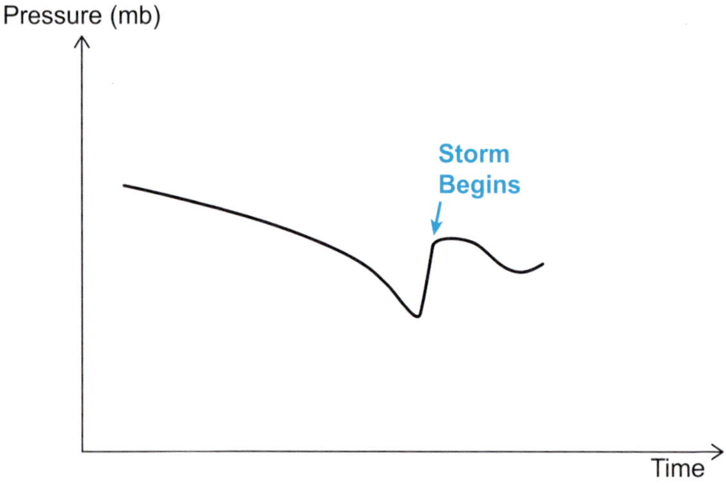

Figure 3. Storm will begin when the short steep rise or jump in pressure is completed after the pressure dip. The temperature makes a corresponding short steep fall when the storm begins.

Storm Early Warning

When the pressure starts falling at more than 1.0 mb / hour you have an Early Warning that a storm may be approaching.

Slowly Developing Thunderstorms

The barograph trace of some thunderstorms initially shows a slowly falling pressure (say, 1.0 mb / 3h) trend and will not exceed The Thunderstorm Rule's threshold (3 – 3 Rule) for 8 – 12 hours.

Marine Navigation expert, David Burch, suggests that ***an even earlier wind warning occurs when the last 3 hours pressure fall is 2.0 mb or more.*** You now have at least 3 hours and possibly up to 5 hours Early Warning of an ordinary thunderstorm.

Storms that Exceed the Threshold from the Outset

The barograph of a Severe Thunderstorm will show the pressure falls steadily more than 8 mb for at least 8 hours to below 1005 mb. Typically, the pressure fall exceeds *The Thunderstorm Rule* threshold from the outset and wind shear contributes to the storm's longevity.

You have at least 5 hours (and possibly up to 9 hours) ***Early Warning of a Severe Thunderstorm*** when the pressure falls 3.0 mb (or more) in 3 hours (or less) ***without*** any sign of storm onset and there is an ongoing steady pressure trend. It is easy to see on a barograph whether after 8 hours the pressure will eventually fall below 1005 mb.

Wind Warnings

You get a wind warning by monitoring the rate of change of pressure. Marine navigation gives further challenges as a wind speed greater than 15 knots (28 km/h) can capsize a small boat.

A 6 mb rise or fall in barometric pressure over 3 hours will increase the wind speed to 54.2 km/h or 29.3 knots from dead calm. A 10 hPa rise or fall in barometric pressure over 3 hours will produce a gale.

Also, a rise or fall in barometric pressure of more than 3 mb per hour will produce a gale at sea. Land winds are about 2/3 of the offshore wind strength.

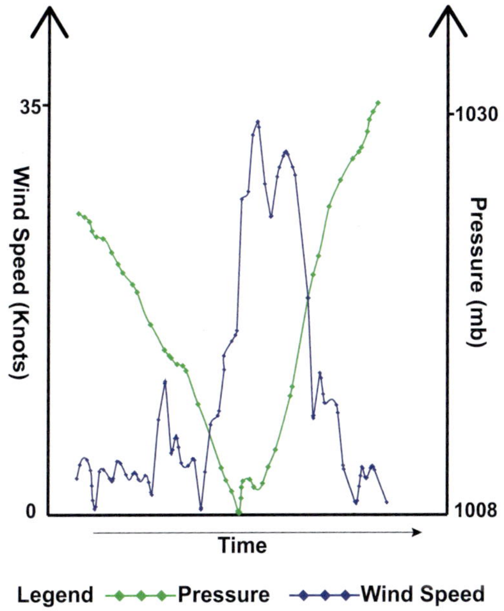

Figure 4. Storm Wind Speed and Pressure Traces

Steady pressure in tropics is hurricane harbinger

Cyclones/hurricanes/typhoons occur in the tropics in late summer and autumn. There is a twice daily rise and fall of pressure on Earth, which is more marked in the tropics. It peaks around 10 am and 10 pm. In the absence of a hurricane, the barometric pressure in the tropics varies about 3 mb each day. At least 30 hours before an approaching hurricane making landfall, the pressure becomes almost steady for 6 hours.
This 6-hour steady pressure interval that occurs before the cyclone's pressure dip is an important early warning sign of a hurricane's development. It is the **Hurricane Harbinger.**

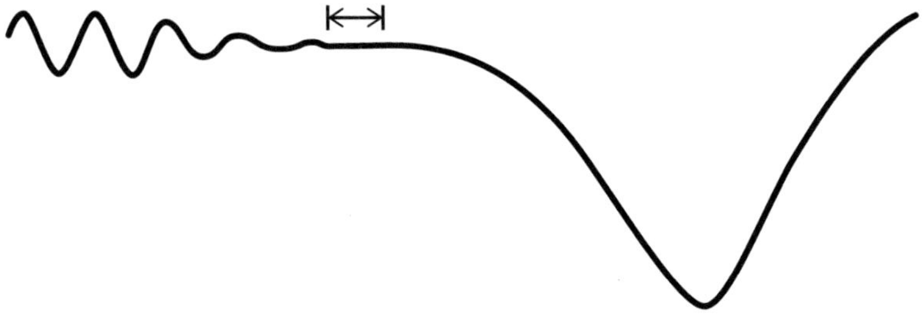

Figure 5: Stylised barogram showing 6-hour steady pressure interval.

The steady pressure interval, after which the pressure dips, is unique for each cyclone and its value is near the seasonal average pressure (e.g. 1016 mb or 30 in). The 6-hour steady pressure interval for a Category 3 cyclone (Category 2 hurricane) commences at least 30 hours before landfall. See Tropical Cyclone Classification at: http://www.srh.noaa.gov/jetstream/tropics/tc_classification.html).
Even if the pressure curve is not entirely flat for the last 6 hours, significant flattening will still be apparent from the barograph, giving you early warning.

Hurricane Early Warning Rule

24-hours early warning of a hurricane is when the pressure has been almost steady for 6 hours.

Remember that the Hurricane Early Warning Rule also applies to cyclones and typhoons.

Figure 6: *Hurricane viewed from satellite.*

Some visual features of Supercell Thunderstorms

A Supercell Thunderstorm has some visual features or aspects you can use to identify it:

1. Overshooting cloud

The anvil is a flat cloud at the top. A strong updraft will make an overshoot (bubble) of cloud on top of the anvil. If this bubble persists for over 10 minutes, it is a sign of a Supercell Thunderstorm.

Figure 7. *Overshooting top cumulonimbus cloud.*

2. Rain-free base

Figure 8. *Rain shaft left, rain-free base of supercell thunderstorm right. Key West, Florida, U.S.*

3. Wall Cloud

If there is a separate cloud below both the rain-free base and the main updraft, it is called a wall cloud or pedestal cloud. It is a sign of a Supercell Thunderstorm.
The wall cloud is located beneath the dark updraft. The wall cloud can show streaks of cloud that suggest rotation.

Figure 9. *Ominous wall cloud portending possible violent weather. Nebraska, USA.*

Spotting Downbursts

There are several visual signs that a downburst is either underway or about to occur.

Virga: Precipitation streaks from the cloud, but does not reach the ground. The atmosphere below the clouds tends to be very dry and rainfall evaporates before it touches the ground. Gusty winds occur in the area of the virga.

Figure10. *Virga with a dry microburst. Photo by Brian Morganti.*

Dust foot: A plume of dust/dirt that is raised as the downburst reaches the ground and moves away from the impact point.

Figure 11. *Dust foot/microburst. Photo by Brian Morganti.*

Rain foot: The rain foot is a pronounced outward bend of the precipitation area near the ground, marking an area of strong outflow winds.

Figure 12. *A wet microburst. Photo by Jim LaDue.*

Figure 13. *Wet downburst with rain foot on the left of rain shaft area. Photo by Brian Morganti*

Storm in Nearby Region

If *The Thunderstorm Rule* is not fully met but the barograph curve still looks like the classic storm shape with a pressure dip and rise in its wake, it means that the storm will **not** occur at your **immediate location** but in a nearby region or suburb.

Use of Drones

You can see the features of an approaching storm earlier if you use a drone with a camera. Drones allow you to see over mountains and confirm storm features such as rotation or if a funnel is developing. A drone can also do an aerial survey to assess the extent of damage after a storm has passed.

Drones can see far from their vantage point above the tree line. They can now be quickly deployed with operators requiring a minimal technical background. There are a range of low cost drones available.

Figure 14. *Tornado damage captured by uncrewed aircraft*

PREDICTING A THUNDERSTORM A WEEK AHEAD

Quick Start Now

Go to https://barometricpressure.app → Select button Use my location → Scroll down the screen and select button

Note: can be viewed on a laptop, PC or iPad.

7 days

Move your cursor along the barometric pressure curve until you reach the lowest point on the forecast:

Note down the time and date of lowest point (spike down minimum) if below 1009 mb as shown on the curve highlighted by your cursor.

The time needs to be in the afternoon or evening for storms on land. If the pressure is above 1009 mb, there can be no storms.

Storm: evening Saturday 28 Dec.

Storm fundamentals

The sea level pressure at the outermost boundary of a thunderstorm is 1009 mb. A thunderstorm cannot form when the pressure remains above 1009 mb.

Even when the pressure is below 1009 mb, no storms can occur if the pressure is steady or rising, or if falling pressure has not reached the overall minimum. It requires a storm or passing cold front (rare in Storm Season) for the pressure to suddenly go from rapidly falling to rapidly rising These fundamentals are based on the laws of physics on which we all rely.

Figure 15. *Thunderstorm pressure*

If the pressure remains above 1009 mb, there can be NO storms.

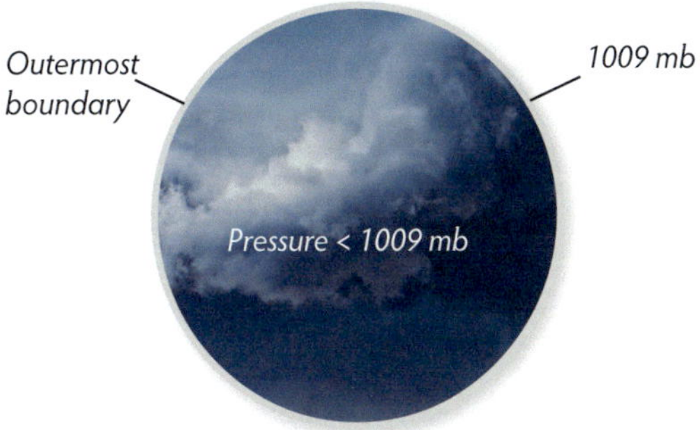

Figure 16. *No storms possible if pressure remains above 1009 mb.*

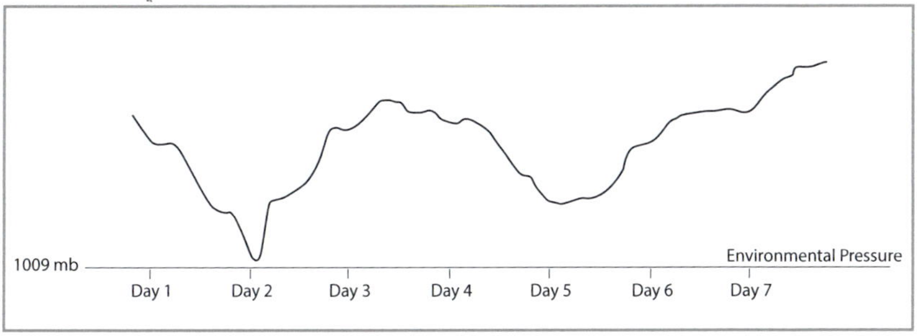

You will also now be able to predict with certainty when there will be no storms. This alone will keep you in good stead for the rest of your life!

You are now able to predict when a storm will occur in the next 7 days at home or on your holiday.

Predicting a Severe Thunderstorm

You can predict a Severe Thunderstorm a week ahead from the pressure curve of the 7-day forecast as follows:

Look at the lowest point on the curve of the 7-day forecast. If the pressure has fallen steadily (straight-line) to that point by more than 8.0 mb in 8 hours (or less) to below 1005 mb, a Severe Thunderstorm is predicted. This is an alternative expression of the rule for a Severe Thunderstorm.

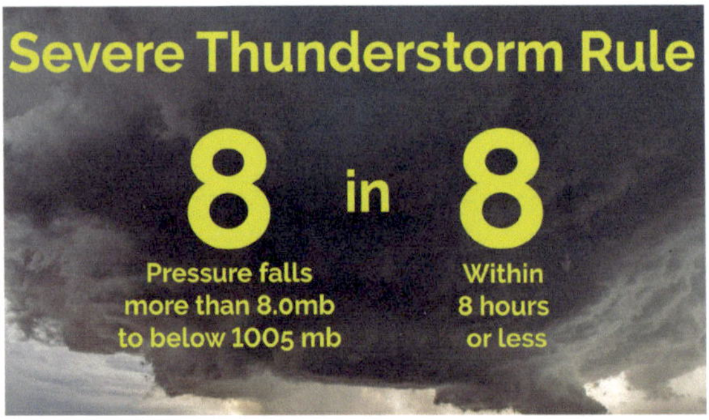

Predicting a Thunderstorm Embedded in a Cold Front

Cold Fronts are troughs of low pressure. As a front approaches, you are moving into an area of lower pressure. A faster, more intense front (a "crashing " front) will show a steeper, more dramatic V-Shape or U-shape barograph. While the pressure drops, expect warming temperatures.

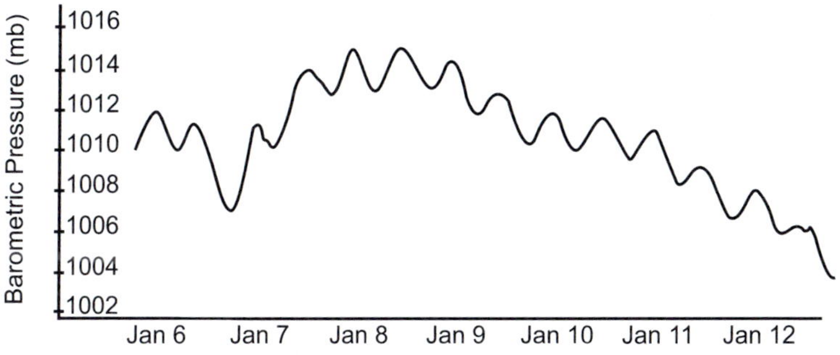

Figure 17. *Barograph of Cold Front passing*

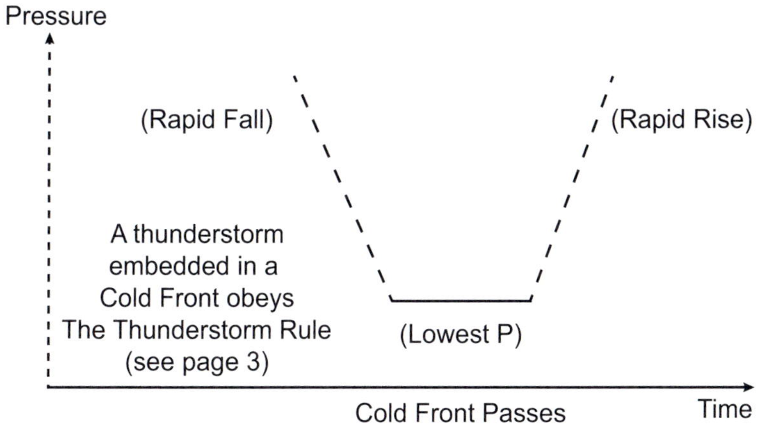

Figure 18. *Pressure pattern when Cold Front passes*

Figure 19: *A cold front moving in at sunset*

You can predict a Thunderstorm embedded in a Cold Front a week ahead from the pressure curve of the 7-day forecast:

Go to barometricpressure.app website (see more on p.19). Move your cursor down the steady pressure curve from the top and observe the pressure reading each hour. When the pressure starts falling at more than 1.0 mb / hour you have early warning that a storm may be approaching. Confirm it when the last 3 hours pressure fall is 3.0 mb or more.

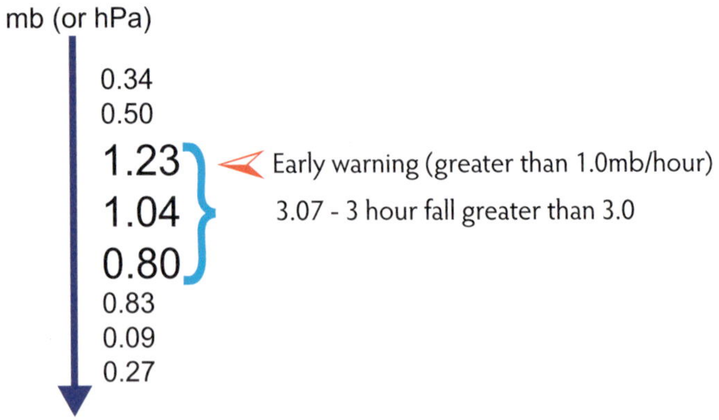

Figure 20: *Example of Hourly Pressure Fall for an ordinary Thunderstorm Thunderstorm in Sydney, Australia, Sunday 22 February 2026*

Find More About Predicting Storms

Predicting Storms – The Adventure Begins,
Third Edition 2024

Available: bestlittlebookshopintown.com

Storm Early Warning rules for different types of storms are given.

Covered: Severe Thunderstorms, tornadic supercell thunderstorms, cyclones, hurricanes, typhoons, extratropical cyclones, tropical storms, tornadoes, firestorms, fire tornadoes, weather bombs, snowstorms, windstorms and dust storms.

References:

A.D. Del Genio, A. D., Yao, M.-S., and Jonas, J., (2007)
Will moist convection be stronger in a warmer climate?,
Geophys. Res. Letters, vol. 34, L16703,
doi:10.1029/2007GL030525, page 2.

Burch, David (2009),
The Barometer Handbook, Starpath, p98.

Ellis, Robert (2024)
Predicting Storms – The Adventure Begins,
Third Edition, Goldener-Parnell Publishing pp.21, 107-8.

Copyright Permissions:

Figure 1 – Cirrus at sea. Credit: Photographer Lieutenant Elizabeth Crapo, NOAA Co, and National Oceanic and Atmospheric Administration/Department of Commerce. Source: Image ID wea03609, NOAA National Weather Service Collection.

Figure 6 – Hurricane viewed from satellite. Credit and source: https://pixabay.com/en/hurricane-earth-satellite-tracking-92968/ (public domain)

Figure 7 – Overshooting top cumulonimbus cloud. Credit: National Oceanic and Atmospheric Administration/ Department of Commerce.
Source: http://www.photolib.noaa.gov/htmls/wea00106.htm.

Figure 8 – Rain shaft to left, rain-free base of severe thunderstorm to right. Key West, Florida, U.S. Credit: Photographer Lieutenant Debora Barr NOAA Corps, and National Oceanic and Atmospheric Administration/Department of Commerce. Source: Image ID wea02191, NOAA National Weather Service Collection.

Figure 9 – Ominous wall cloud portending possible violent weather.Image ID: wea04209, NOAA's National Weather Service (NWS) Collection. Location: Nebraska, near Atkinson.
Photo Date: 2010 May 31.
Photographer: Jerry Penry, Registered Land Surveyor.

Figures 10,11, 12,13 – Words and images from Spotters Guide 2011 - Fielduse. https://www.weather.gov/media/unr/SpottersGuide2011_fielduse.pdf. NOAA's National Weather Service (NWS).

Figure 14 –Tornado damage captured by uncrewed aircraft-Sawyerville-AL-27March2021-NOAA Research-landscape.jpg

Figure 19 – NOAA Weather in Focus Photo Contest March 2015 Photographer Kenneth Flowers, Location New Mexico, Tatum.

All other photos are licensed.

About the Author

Robert Ellis has a BSc (Hons) from Sydney University and has completed postgraduate qualifications at three Australian universities. He has published internationally in the US.

He served for seven years as a Flight Lieutenant with a Permanent Commission in the RAAF. He taught aircrew at the Central Flying School and the School of Air Navigation and taught on the first weapons systems course in Australia. He was a project manager in Technical and Further Education for 25 years.

Robert is a scientist and storm expert who has been referred to in recent years as a storm chaser in the Press.

My Motto

Life is only froth and bubble
but two things stand alone
kindness in another's troubles
and courage in your own.

Made in the USA
Las Vegas, NV
09 November 2026

57757228R00021